零失敗
秘方系列

U0100077

啖啖鮮香
住家餸

The best of homestyle cooking

目錄

Contents

健康烹飪
Healthy cooking

　　許多上班一族都會外出用午膳，晚上又約會朋友晚餐。經常外食，會令身體吸收太多油脂，進食太多味精，味蕾變得遲鈍了，口味越來越重。

　　為了身體健康，多點在家煮飯吧！

　　住家餸一點也不單調，視乎你是否懂得煮、懂得材料的配搭。這書介紹多款健康烹調法如下：

　　蒸：只要調味及食材配搭得宜，少油蒸餸也令人添飯！

　　水炒：以水代油炒煮材料，吃得清爽；但緊記用少許油爆香料頭，自然提升美味指數。

　　水煮：挑選香味濃郁的配菜同煮，或用其浸泡的汁液精華炮製，以食材原味煮成靚餸。

　　烤焗：烘焙讓食材的油脂溢出，味道濃郁，香口引人，嘗到材料之原汁原味！

　　此外，還介紹健康凍食和燜餸。

　　在這裏分享一些少油烹調的技巧，讓各位煮得更得心應手，吃得健康：

　　建議選用易潔鑊烹調，用油量比一般煮食用具少，減少食材吸收油分的機會，但緊記易潔鑊不能用大火長時間爆炒。

　　用洋葱、乾葱、薑、蒜等料頭起鑊爆香，香味濃郁，增添餸菜的香味，彌補少油炒煮不足之香氣。

　　少用醬料炒煮，因醬料需要多油量爆炒才能散發香味。

　　選用煮食用的橄欖油或菜油，對身體有益。市面有噴霧式的橄欖油發售，用油量更少。

　　選購脂肪含量少的肉類，如柳梅、梅頭、豬腱、瘦腩肉或雞肉等。

　　多選用香味濃郁的食材配搭，如蝦乾、蝦米、沖菜、土魷、蜆肉或芫茜等，令餸菜香氣四溢！

Most office workers eat out at lunch and eat out again with friends at dinner. Eating out too frequently would accumulate too much fat in the body, not to mention the excessive intake of MSG and the overpowering seasoning. You'd find your taste buds getting dull and you'd crave food with even stronger flavours.

For your health, just stay in and make yourself dinner at home.

A homestyle meal is nothing but boring – it all depends on how skilful you are as a chef and whether you can come up with interesting food combinations. This book covers all grounds of healthy cooking, including:

Steaming: If you use the right seasoning with the right ingredients, you can turn out low-fat steamed dishes so delicious that your guests would request a refill of rice.

Stir-frying with water: You can stir-fry ingredients in water instead of oil to make it light and less greasy. Just make sure you stir-fry the aromatics in a dash of oil before adding other ingredients. That's enough to elevate the aroma and flavours in major ways.

Boiling in water: Pair bland-tasting ingredients with flavoursome condiments and cook with in a little water. Or, use the water that flavoursome ingredients are soaked in as a sauce. You can then make a mouth-watering dish with their authentic flavours coming through.

Baking / grilling: The oil in the ingredients will drain when grilled and flavours will turn more intense with a spellbinding smoky aroma. The authentic flavours of the ingredients will shine in full glory.

This book also covers the preparation of cold dishes and stews.

Notes for Less Oil Cooking

Unlike other cooking utensils, non-stick pans require less oil in cooking food. You are assured of a healthier diet by taking food that absorbs less oil. Non-stick pans cannot be used over high heat for stir-frying for a long time.

Food cooked with less oil does not give a fragrant smell. To enhance the aromatic flavour, stir-fry the sweet-scented spices like onion, shallot, ginger and garlic before adding other ingredients.

Use less sauce in cooking as it needs to be fried with a lot of oil to help spread its fragrance.

Choose healthy cooking oil like olive oil or vegetable oil. Available in the market, olive oil spray helps minimize the use of oil in cooking.

Buy meat such as tenderloin, pork butt, pork shin, lean pork belly or chicken. They all contain less fat.

Often use different mixtures of food ingredients with strong aromatic flavour to make dishes smelling great! Examples are dried prawn, dried shrimp, preserved turnip, dried squid, shelled clam, coriander, and more.

白 切 雞
Steamed Fresh Chicken with Spring Onion

材料

鮮活清遠雞 1 隻（約 3 斤）
葱 6 條
薑汁半湯匙
粗鹽 1 湯匙

蘸汁

上等生抽 1 小碟

做法

1. 清遠雞劏好，雞嘴及雞尾用粗鹽擦淨，再用水洗淨，瀝乾水分。

2. 雞腎刮淨內膜及脂肪，用粗鹽擦淨，再用水洗淨；雞肝去掉脂肪，用水洗淨。

3. 葱去掉鬚頭，洗淨，切長段。

4. 雞內腔用粗鹽及薑汁塗勻，置半小時。

5. 將半份葱段排在蒸碟底，放上鮮雞（雞胸向上），鋪上餘下之葱段，與雞內臟一併隔水用大火蒸 25 分鐘，關火，加蓋焗 15 分鐘。

6. 預備半鍋盛有冰粒之凍開水，原隻雞放入冰水內浸凍，瀝乾水分，斬件上碟，伴以蘸汁供食。

Ingredients

1 live Qing Yuan chicken (about 1.8 kg)
6 sprigs spring onion
1/2 tbsp ginger juice
1 tbsp coarse salt

Dipping sauce

1 small dish premium light soy sauce

Method

1. Slaughter the chicken. Rub the beak and buttock with the coarse salt. Rinse and drain.

2. Scrape the membrane and fat off the chicken gizzard. Rub with the coarse salt and rinse. Remove fat from the liver. Rinse.

3. Cut away the root of the spring onion. Rinse and cut into long sections.

4. Spread the coarse salt and ginger juice on the chicken cavity. Rest for 1/2 hour.

5. Place 1/2 portion of spring onion on a plate. Put the chicken on top (breast upward). Arrange the rest of the spring onion on the chicken. Put the internal organs of the chicken on the side of the plate. Steam over high heat for 25 minutes. Turn off heat. Leave for 15 minutes with a lid on.

6. Prepare 1/2 pot of cold drinking water with ice cubes. Soak the whole chicken into the water to allow it to cool down. Drain. Chop into pieces. Serve with the dipping sauce.

◯◯ 零失敗技巧 ◯◯
Successful Cooking Skills

為何選吃清遠雞？

清遠雞脂肪分佈平均，肉滑鮮美、外皮爽脆，而且飼養時自由走動，肉質結實，原隻蒸吃可嘗其原汁原味。

Why choose Qing Yuan chicken?

The free-range Qing Yuan chicken has a firm meat texture with even distribution of fat. Steamed whole chicken gives the original flavour with smooth meat and crispy skin.

用半湯匙薑汁醃雞，足夠嗎？

薑汁帶濃郁的香辛味，半湯匙的份量塗抹雞內腔，已令鮮雞香氣四溢！

Is it enough to marinate the chicken with 1/2 tbsp of ginger juice?

The pungent ginger juice makes the chicken very fragrant by just spreading 1/2 tbsp of the juice on the chicken cavity!

怎樣才會令雞皮爽肉滑？

原隻雞蒸熟後，立即放入盛有冰粒之凍開水浸凍，因冷縮熱脹的關係，皮爽肉滑，滋味無窮！

How do you keep the chicken skin springy and the flesh silky?

After you steam the whole chicken, put it into a big bowl of drinking water with ice cubes. The ice water bath would shrink the skin and stop the cooking process instantly, making the skin springy and the flesh juicy and tender.

此蒸雞法與平常慣做的方法，食味有何不同？

雞肉不帶血腥味，皮爽、肉質結實，味鮮香！

How does the taste differ between this steaming method and regular cooking?

The steamed chicken has no bloody smell. The meat is firm and the skin is crunchy. Yummy!

醉花蟹

Drunken Coral Crab with Shaoxing Wine

◎ 材料

花蟹 1 隻（約 12 兩）
薑 3 片
葱 4 條
紹酒 250 毫升

◎ 調味料

泰國魚露 250 毫升
清水 400 毫升
糖 6 湯匙

◎ 做法

1. 煮滾調味料，待涼，加入紹酒拌成汁料。
2. 花蟹去除內臟，刷洗乾淨，瀝乾水分。
3. 花蟹、薑及葱同放碟內，隔水蒸約 14 分鐘至蟹全熟，待涼。
4. 蟹斬件，略拍蟹鉗，放入汁料內浸泡，冷藏 2 小時即可享用。

醉
花
蟹

◎ Ingredients

1 coral crab (about 450 g)
3 slices ginger
4 sprigs spring onion
250 ml Shaoxing wine

◎ Seasoning

250 ml Thai fish sauce
400 ml water
6 tbsp sugar

◎ Method

1. Bring seasoning to the boil and set aside to let cool. Add Shaoxing wine and mix to become the sauce.
2. Remove entrails from coral crab. Rub and wash carefully. Rinse and drain.
3. Put coral crab, ginger and spring onion into a plate. Steam for about 14 minutes until done. Set aside to let cool.
4. Chop the crab into pieces. Pat the pincers briefly. Soak them into the sauce. Refrigerate for 2 hours and serve.

◎ 零失敗技巧 ◎
Successful Cooking Skills

汁料內為何加入魚露？

更能提升蟹肉之鮮味，成為薰香醉人之冷盤，令人百吃不厭！

Why add fish sauce into the sauce for soaking crab?

Fish sauce can further enhance the taste of crabmeat and makes this cold dish always tastes good!

除花蟹外，還可選用哪類蟹？

肉蟹也適合製成醉蟹，蟹肉滲滿酒香味！

What kind of crab can be used except coral crab?

Mud crab can also be used.

花蟹買回來不立即炮製，如何養活？

將花蟹（連繩）放於碟上，用濕布蓋着，可養活大半天。

How to raise coral crab if not cooked immediately after bought?

Put the crab (with strings) into a plate and cover with a damp cloth. It can live for more than half day.

椰皇蟹肉鉢仔蛋
Steamed Eggs and Coconut Juice with Crabmeat

蒸

材料
鮮蟹肉 2 兩
椰皇 1 個
雞蛋 4 個
芫茜適量

調味料
鹽半茶匙

用具
瓦缽 1 個（18 厘米）

做法
1. 刺穿椰皇頂部之洞，傾出 250 毫升椰水，備用。
2. 雞蛋及調味料拂勻，與椰水拌勻，傾入瓦缽內，用中大火隔水蒸 4 分鐘。
3. 蛋面鋪上鮮蟹肉，再蒸 3 分鐘，加入芫茜，加蓋稍焗即可。

Ingredients
75 g fresh crabmeat
1 king coconut
4 eggs
coriander

Seasoning
1/2 tsp salt

Tool
1 clay saucer bowl (18 cm)

Method
1. Pinch through the hole at the top part of king coconut. Pour out 250 ml of coconut juice and set aside.
2. Whisk eggs with seasoning. Mix eggs with coconut juice and pour into a saucer bowl. Steam over medium-high heat for 4 minutes.
3. Lay fresh crabmeat over the surface of eggs and steam for 3 minutes more. Put in coriander and cover the lid for a while. Serve.

椰皇蟹肉缽仔蛋

◎ 零失敗技巧 ◎
Successful Cooking Skills

用瓦缽蒸蛋有何技巧？
由於瓦缽傳熱速度較慢，因此蒸首 4 分鐘時，建議使用中大火。
What's the technique of steaming eggs in saucer bowl?
Since clay saucer bowl is slow in heat transmission, it is suggested to use medium-high heat at the first 4 minutes of steaming.

蒸蛋與椰水混和，食味如何？
蛋內散發陣陣清新的椰皇香氣，配上鮮甜的蟹肉，確是另一番味覺享受！
What's the taste of steaming eggs together with coconut juice?
The coconut aroma gets into the eggs thus it is a great experience to serve it with fresh and sweet crabmeat.

甚麼是此餸成功之要訣？
必須將椰水及雞蛋完全拌勻，否則難以蒸熟。
What's the hint to make this dish successfully?
The coconut juice and eggs must be mixed well completely or else it is hard to be steamed until done.

椰皇蟹肉缽仔蛋

蒸魚飯
Steamed Fish in Bamboo Basket

◎ 材料
木棉魚（大眼雞）2 條
釘公魚 2 條
椰菜葉 4 塊
粗鹽 1 湯匙

◎ 蘸汁
普寧豆醬 1 小碟

◎ 做法
1. 木棉魚及釘公魚劏好，洗淨，抹乾水分，均勻地抹上粗鹽，醃半小時。
2. 椰菜葉洗淨，抹乾水分，鋪在竹笒箕內，排上木棉魚及釘公魚，隔水大火蒸 10 分鐘，原個笒箕取出，待涼。（圖 1-3）
3. 將魚的外皮撕去，上碟，蘸汁供食。

蒸
魚
飯

◎ Ingredients
2 bigeye fishes
2 four lined terapons
4 cabbage leaves
1 tbsp coarse salt

◎ Dipping sauce
1 small dish Puning bean sauce

◎ Method
1. Gill the bigeye fishes and fourlined terapons. Rinse and wipe them dry. Evenly rub with the coarse salt. Marinate for 1/2 hour.
2. Rinse the cabbage leaves. Wipe them dry. Place in a bamboo basket. Arrange the bigeye fishes and fourlined terapons on top. Steam the fish over high heat for 10 minutes. Take out the basket. Allow it to cool down. (fig. 1-3)
3. Tear the skin off the fish. Serve with the dipping sauce.

1

2

3

◎◎ 零失敗技巧 ◎◎
Successful Cooking Skills

要選哪些鮮魚呢？

蒸魚飯要選購多肉、少幼骨的海魚，如紅鰽、黃花魚、馬友及池魚，可嘗到啖啖魚飯鮮味啊！

Which type of fish should we choose?

Select meaty marine fish with less tiny bones, such as red seabream, yellow croaker, threadfin and Japanese jack mackerel. They can be made into fresh fish rice.

為甚麼不需要刮去魚鱗？

用作烹調蒸魚飯的鮮魚，不要去掉魚鱗，以保存更多魚的鮮味。

Why didn't you scale the fish?

For this recipe, it's advisable to keep the scales on, so as to preserve the full flavours of the fish.

蒸魚時，為何墊上椰菜葉？

以免水大滾時，熱水浸煮魚肉，令魚的鮮味流失。除椰菜外，生菜、紹菜及黃芽白亦可。

Why place the fish on pieces of cabbage leaves for steaming?

This is to prevent boiling water from overflowing the fish to make the flavour gone. Lettuce, Tianjian cabbage or Napa cabbage can also be used.

金不換蒸鮮帶子
Steamed Fresh Scallops with Basil Leaves

材料

鮮薄殼帶子 8 隻
金不換 1 棵
蒜茸 1 1/2 湯匙
乾葱茸 1 湯匙
紅辣椒絲 1 湯匙

調味料

生抽 1 湯匙
老抽 1 湯匙
糖半湯匙
麻油 1 茶匙
水半湯匙

做法

1. 金不換摘出葉片，洗淨備用。
2. 用刀打開帶子外殼，去腸及去內臟，修剪帶子薄殼，洗淨，瀝乾水分，排於碟上。（圖 1-4）
3. 熱鑊下油，爆香蒜茸、乾葱茸及紅椒絲，待涼，鋪在帶子上。
4. 隔水蒸約 3 至 4 分鐘，灑入金不換葉，稍焗片刻。
5. 分別煮滾少許生油及調味料，依次澆在帶子上，趁熱品嘗。

Ingredients

8 fresh thin-shelled scallops
1 stalk basil leaves
1 1/2 tbsp minced garlic
1 tbsp minced shallot
1 tbsp red chili shreds

Seasoning

1 tbsp light soy sauce
1 tbsp dark soy sauce
1/2 tbsp sugar
1 tsp sesame oil
1/2 tbsp water

Method

1. Tear off leaves from basil. Rinse and set aside.
2. Open the shells of scallops with a knife. Remove entrails. Trim their thin shells. Rinse and drain. Arrange scallops into a plate. (fig. 1-4)
3. Add oil into a hot wok. Stir-fry minced garlic, minced shallot and shredded red chili until fragrant. Set aside to let cool and arrange onto the scallops.
4. Steam for about 3 to 4 minutes. Sprinkle over basil leaves and cover the lid for a while.
5. Bring a little oil and seasoning to the boil respectively. Pour them over the scallops in order. Serve hot.

1

2

3

4

◯◯ 零失敗技巧 ◯◯
Successful Cooking Skills

最後放入金不換葉，香氣足夠滲入帶子嗎？
金不換不宜久煮，故最後放入稍焗即散發芳香之味。

Can the smell of basil leaves go into the scallops if putting in at the last?

Basil leaves cannot be cooked for a long time. Thus putting in at the last step and cover for a while can bring out their aroma already.

如何修剪帶子薄殼？
沿帶子適度修剪，排碟及食用更方便。

How to trim the thin shells of scallops?

Trim around the shells so that they are easier for arranging into the plate and more convenient for consumption.

哪裏購買金不換？
泰國雜貨店或部份菜檔均有出售。

Where to buy basil leaves?

They can be bought from Thai grocery stores and some vegetable stalls.

原味蒸乾三鮮

Steamed Dried Seafood in Three Delicacies

◯◯ 材料
乾吊筒 2 兩
蝦乾 1 兩
銀魚乾 1 兩
薑絲 1 湯匙

◯◯ 調味料
上等生抽半湯匙
熟油 3/4 湯匙

原味蒸乾三鮮

◯◯ 做法
1. 乾吊筒、蝦乾及銀魚乾分別洗淨，備用。
2. 乾三鮮放入蒸碟內，排上薑絲，隔水大火蒸 10 分鐘，取出，澆上調味料即成。

◯◯ Ingredients
75 g dried squid
38 g dried prawns
38 g dried anchovies
1 tbsp shredded ginger

◯◯ Seasoning
1/2 tbsp premium light soy sauce
3/4 tbsp cooked oil

◯◯ Method
1. Rinse the dried squid, dried prawns and dried anchovies respectively. Set aside.
2. Put all the dried seafood on a plate. Arrange the ginger on top. Steam over high heat for 10 minutes. Remove. Drizzle with the seasoning to finish.

◎ 零失敗技巧 ◎
Successful Cooking Skills

乾三鮮毋須用水浸泡嗎？
毋須用水浸泡，以免流失原有之鮮味，只需洗淨即可蒸煮。
Need not soak the dried seafood in water?
Just rinsing them will do to avoid losing their raw flavour.

如何挑選乾三鮮？
建議選香味充足、乾爽、帶光澤及原整的乾三鮮。
How to choose the dried seafood?
Good dried seafood is fragrant, dry, glossy and intact.

瑤柱三文魚醬蒸釀豆腐

Steamed Stuffed Beancurd with Pork, Flaked Salted Salmon and Dried Scallops

(◯) **材料**

布包豆腐 2 塊
免治豬肉 3 兩
瑤柱三文魚醬 1 1/2 茶匙
葱粒半湯匙
雞蛋 1/4 個
粟粉 1 茶匙
芫茜少許

(◯) **調味料**

生抽 2 茶匙
熟油半湯匙

(◯) **做法**

1. 免治豬肉、葱粒、雞蛋、半份瑤柱三文魚醬及粟粉拌成餡料。

2. 布包豆腐切塊，鋪上適量餡料，排於蒸碟內，綴上餘下之瑤柱三文魚醬，隔水大火蒸 8 分鐘，隔去汁液，澆上調味料，以芫茜葉裝飾即成。

(◯) **Ingredients**

2 cubes cloth-wrapped beancurd
113 g ground pork
1 1/2 tsp flaked salted salmon with dried scallops
1/2 tbsp diced spring onion
1/4 egg
1 tsp cornstarch
coriander

(◯) **Seasoning**

2 tsps light soy sauce
1/2 tbsp cooked oil

(◯) **Method**

1. Mix ground pork, diced spring onion, egg, half of the flaked salted salmon with dried scallops, and cornstarch. This is the filling.

2. Cut the tofu into pieces. Put some filling on each piece. Arrange the stuffed beancurd in a steaming plate. Then divide the remaining flaked salted salmon with dried scallops among the stuff beancurd by putting it on top. Steam over high heat for 8 minutes. Drain any liquid on the plate. Pour the seasoning over. Garnish with coriander. Serve.

瑤柱三文魚醬

Flaked Salted Salmon with Dried Scallops

材料 of 材料

急凍三文魚 1 塊（約 300 克 / 半斤）
瑤柱半兩
薑茸 2 湯匙
蒜茸 1 湯匙
米酒 2 湯匙
粟米油 1 1/2 杯

做法 做法

1. 三文魚解凍，洗淨，抹乾水分。

2. 取瓦瓷或塑膠容器，先鋪一層粗海鹽，放入魚扒，最後蓋滿粗海鹽。

3. 加蓋，密封，冷藏於雪櫃的生果格層 2 至 3 星期成鹹三文魚。

4. 瑤柱洗淨，用水浸軟，撕成瑤柱絲。

5. 鹹三文魚隔水蒸 10 分鐘，待冷，拆肉弄散。

6. 燒熱鑊下油半杯，下瑤柱絲炒香，加入薑茸炒香，下三文魚肉及蒜茸用慢火炒透及壓碎，潛酒，邊炒邊加入餘下的油（油必須蓋過所有材料），再炒片刻，盛起，待冷入瓶儲存。

Ingredients of Ingredients

300 g frozen salmon
19 g dried scallops
2 tbsp grated ginger
1 tbsp grated garlic
2 tbsp rice wine
1 1/2 cups corn oil

Method of Method

1. Thaw 1 piece of frozen salmon. Rinse well and wipe dry.

2. Line a ceramic or plastic container with a layer of coarse sea salt. Put in the salmon and top with more sea salt until full.

3. Cover the lid and seal well. Store in the crisper compartment of your fridge for 2 to 3 weeks.

4. Rinse the dried scallops. Soak them in water until soft. Tear them apart into fine shreds.

5. Steam the salted salmon for 10 minutes. Leave it to cool. Skin and de-bone it. Break it down into flakes.

6. Heat a wok and add 1/2 cup of oil. Put in the dried scallops and stir until fragrant. Add grated ginger and stir fry until fragrant. Add flaked salmon and garlic. Stir fry until done and crush well. Sizzle with wine. Pour in the remaining oil while stirring continuously. There must be enough oil to cover all solid ingredients. Stir fry briefly. Set aside to let cool. Transfer into sterilized bottles.

瑤柱三文魚醬蒸釀豆腐

瑤柱三文魚醬

用粗鹽醃三文魚

◎◎ 零失敗技巧 ◎◎
Successful Cooking Skills

為何用布包豆腐？
包布豆腐的質感嫩滑，不容易散爛。

Why cloth-wrapped beancurd?

Because it is soft and tender, yet without being crumbled easily.

瑤柱三文魚醬為何可儲存半年？
由於三文魚已用鹽醃味，而且醬料用油密封，故能保存鮮味！

How come the flaked salted salmon with dried scallops can last up to 6 months?

It's because the salmon has been salted and the oil seals it off from getting in touch with air. That's why it could last that long.

炮製瑤柱三文魚醬有甚麼竅門？
瑤柱及三文魚必須用油炒至乾透，徹底去掉所有水分，令醬汁更美味，而且耐存。

Is there any trick in making dried scallop salmon sauce?

Both the dried scallops and the salmon have to be fried in oil till dry. By removing all moisture in the both ingredients, the sauce would have a robust intense flavour. That also extends its shelf life.

豆腐蛋白蒸鮮蜆
Steamed Clams with Beancurd and Egg Whites

材料
布包豆腐 1 塊
鮮蜆 4 兩
蛋白 1 個
粟粉 1 茶匙
芫茜少許

醃料
胡椒粉少許
鹽 1/3 茶匙
水半杯

調味料
生抽 2 茶匙
熟油 2 茶匙

做法
1. 用叉子搓爛布包豆腐，加入蛋白、粟粉及醃料拌勻，放入深蒸碟內。

2. 鋪上鮮蜆，隔水用大火蒸 15 分鐘，取出，澆上調味料，以芫茜裝飾即可。

Ingredients
1 cloth-wrapped beancurd
150 g fresh clams
1 egg white
1 tsp cornflour
coriander

Marinade:
ground white pepper
1/3 tsp salt
1/2 cup water

Seasoning
2 tsp light soy sauce
2 tsp boiled oil

Method
1. Mash the beancurd with a fork. Mix well with the egg white, cornflour and the seasoning. Put on a deep plate for steaming.

2. Place the clams evenly on top. Steam over high heat for 15 minutes. Take out. Drizzle with the seasoning. Garnish with the coriander and serve.

◎ 零失敗技巧 ◎
Successful Cooking Skills

鮮蜆買回來後，需要特別處理嗎？

大部份於街市出售之鮮蜆，不含沙粒，只要洗擦乾淨外殼即可烹調。

How to treat the fresh clams bought from the market?

Most of the clams sold in the market contain no sands. You only need to rub and wash the shell.

布包豆腐適合用於蒸餸嗎？

是，因布包豆腐質感嫩滑，吃起來滑溜溜。

Is cloth-wrapped beancurd suitable for steaming?

Yes, it is. It has a silky texture and taste smooth after steaming.

應選購哪類鮮蜆烹調？

只要是活蜆，任何品種也可烹調成美味佳餚；建議選用花蛤較佳。

What kind of fresh clams should we choose?

All kinds of live clams can be used to make great delicacies, but Venus clams are much better.

Healthy

鮮竹四寶蔬

Braised Beancurd Stick
with Assorted Vegetables

◎ 材料

鮮腐竹 3 條
本菇 1 包
蜜糖豆約 15 條
小粟米 8 條
小棠菜 4 兩
薑 4 片

◎ 調味料（調勻）

蠔油半湯匙
生抽半湯匙
糖 1/4 茶匙
胡椒粉少許
粟粉 1 茶匙
水 3 湯匙

◎ 做法

1. 本菇切去末端；蜜糖豆撕去老筋；小粟米切斜度，全部洗淨。

2. 鮮腐竹略洗，壓乾水分；小棠菜洗淨，備用。

3. 本菇、小粟米及蜜糖豆同飛水，瀝乾水分。

4. 燒熱鑊，下油 2 湯匙，下薑片炒香，加入小棠菜炒片刻，再下鮮腐竹、本菇、小粟米及蜜糖豆炒勻，最後加入調味料煮 5 分鐘即成。

◎ **Ingredients**

3 fresh beancurd sticks
1 packet Hon-shimeji mushrooms
15 sugar snap peas
8 baby corns
150 g Shanghainese white cabbages
4 slices ginger

◎ **Seasoning (mixed well):**

1/2 tbsp oyster sauce
1/2 tbsp light soy sauce
1/4 tsp sugar, ground white pepper,
ground white pepper
1 tsp cornflour
3 tbsp water

◎ **Method**

1. Remove the root of the mushrooms. Tear off the hard strings of the sugar snap peas. Cut the baby corns diagonally. Rinse all.

2. Slightly rinse the beancurd sticks. Squeeze them dry. Rinse the Shanghainese white cabbage. Set aside.

3. Scald the mushrooms, baby corns and sugar snap peas together. Drain.

4. Heat a wok. Put in 2 tbsp of oil. Stir-fry the ginger until fragrant. Add the Shanghainese white cabbage and stir-fry for a while. Add the beancurd sticks, mushrooms, baby corns and sugar snap peas. Stir-fry and mix well. Put in the seasoning at last and cook for 5 minutes. Serve.

鮮竹四寶蔬

◎ 零失敗技巧 ◎
Successful Cooking Skills

如何撕去蜜糖豆的老筋？
首先撕出蒂部，再沿豆邊撕去老筋即可。
How to tear off the hard strings of sugar snap peas?
Tear off the stalk on top. Then tear off the hard strings along the edge.

最後加入調味料煮 5 分鐘，味道足夠嗎？
全部材料容易熟透及入味，煮 5 分鐘已足夠；但略硬的材料建議煮久一些。
Will it have enough taste by cooking with the seasoning for just 5 minutes?
It is enough as all the ingredients will absorb the seasoning and get done quickly. For tougher ingredients, cook longer.

鮮腐竹買回來後，需要特別處理嗎？
只需原包鮮腐竹放入冰箱冷藏，毋須處理，使用時取出即可。餘下的份量建議即時急凍處理。
How to handle fresh beancurd sticks before using?
Keep the packed beancurd sticks in a refrigerator and take out when use. For the extra ones, it is better to freeze them at once.

陳皮豆豉醬清蒸泥鯭

Steamed Rabbit Fish in Dried Tangerine Peel and Black Bean Sauce

材料

活泥鯭 12 兩
紅椒圈少許
陳皮豆豉醬半湯匙
熟油 1 湯匙
芫茜 2 棵

做法

1. 泥鯭魚劏好，洗淨，排於蒸碟上。

2. 鋪上陳皮豆豉醬及紅椒圈，隔水大火蒸 7 分鐘，隔去部分汁液，澆上熟油，伴上芫茜即成。

Ingredients

450 g live white spotted rabbit fish
red chilli (cut into rings)
1/2 tbsp dried tangerine peel and black bean sauce
1 tbsp cooked oil
2 sprigs coriander

Method

1. Dress the rabbit fish. Rinse well. Arrange on a steaming plate.

2. Spread dried tangerine peel and black bean sauce over the fish. Sprinkle red chilli rings on top. Steam over high heat for 7 minutes. Drain any liquid on the dish. Dribble boiling hot oil on it. Garnish with coriander.

陳皮豆豉醬

Dried Tangerine Peel and Black Bean Sauce

◎ 材料

上等豆豉 2 兩
陳皮 1 個
蒜茸半湯匙
紅椒碎 1 茶匙

◎ 調味料

紹酒 1 湯匙
生抽 1 湯匙
老抽半湯匙
糖半湯匙
胡椒粉少許

◎ 做法

1. 陳皮用水浸軟，刮淨內瓤，切碎；
   豆豉用水沖洗，切碎，備用。

2. 燒熱鑊下油 3 湯匙，下豆豉炒至
   香，加入陳皮、蒜茸、紅椒碎及
   調味料炒至散發香味，待涼，入
   瓶儲存。

◎ Ingredients

75 g premium fermented black beans
1 whole dried tangerine peel
1/2 tbsp grated garlic
1 tsp chopped red chilli

◎ Seasoning

1 tbsp Shaoxing wine
1 tbsp light soy sauce
1/2 tbsp dark soy sauce
1/2 tbsp sugar
ground white pepper

◎ Method

1. Soak dried tangerine peel in water
   until soft. Scrape off the pith. Finely
   chop it. Set aside. Rinse the fermented
   black beans in water. Finely chop
   them. Set aside.

2. Heat a wok and add 3 tbsp of oil. Stir
   fry black beans until fragrant. Add
   dried tangerine peel, grated garlic,
   red chilli and seasoning. Stir fry until
   fragrant. Leave it to cool. Store in
   sterilized bottles.

◎ 零失敗技巧 ◎
Successful Cooking Skills

要怎樣處理陳皮？

陳皮是此醬料的主角。如選用 3 年陳皮，果瓤薄、味香濃的，浸後不需要刮去瓤；如選用 3 年以內的，稱為果皮，浸後需要刮去瓤，去掉苦澀味，才能煮出香濃的醬汁。此醬放於雪櫃可儲存 2 個月。

How should I prepare the dried tangerine peel?

Dried tangerine peel is the key condiment in this sauce. If you use dried tangerine peel which is 3 years old or older, the pith is thin and the citrus fragrance is strong. Just soak it in water till soft without scraping off the pith. However, if you use any dried tangerine peel that is less than 3 years old, you should scrape off the pith after soaking it to remove the bitterness. That's how you turn out a rich and aromatic sauce. This sauce lasts in the fridge for 2 months.

為何陳皮豆豉醬與泥鯭如此搭配？

陳皮的香氣可去掉泥鯭的泥腥味，是絕佳配搭！

Why does rabbit fish match so well with the dried tangerine peel and black bean sauce?

It is because the citrus fragrance of the dried tangerine peel helps remove the muddy taste of rabbit fish. They are the perfect match.

購買深海泥鯭還是普通泥鯭？兩種食味有何不同？

深海泥鯭肉質豐厚，魚味鮮，泥腥味少，但價錢較貴；普通泥鯭肉質鮮嫩，泥腥味重，故須用陳皮豆豉醬辟味。

Should I get regular rabbit fish or deep sea rabbit fish? What are their differences?

Deep sea rabbit fish is fleshy and has a strong seafood flavour. It tastes less muddy than the regular one, but it is considerably more expensive. Regular rabbit fish has tender flesh but also a stronger muddy taste. Thus, you should use dried tangerine peel and black bean sauce to cover it up.

海膽蒸蛋白
Steamed Sea Urchin with Egg White

◎ **材料**
海膽 40 克
蛋白 4 個（200 毫升）
葱花適量

◎ **調味料**
鹽 1/3 茶匙
清雞湯 200 毫升

◎ **做法**

1. 取半份海膽壓成茸；其餘半份留用。

2. 海膽茸、蛋白及調味料拂勻，傾入深碟內。

3. 隔水用中火蒸 3 分鐘，轉小火再蒸約 4 分鐘，排入餘下之海膽，灑上葱花，續蒸片刻即成。

◎ **Ingredients**
40 g sea urchin
4 egg whites (200 ml)
diced spring onion

◎ **Seasoning**
1/3 tsp salt
200 ml chicken broth

◎ **Method**

1. Crush half portion of sea urchin and set aside the other half.

2. Whisk egg white with crushed sea urchin and seasoning. Mix well and pour into a deep plate.

3. Steam over medium heat for 3 minutes. Turn to low heat and steam for about 4 minutes. Arrange in the remaining sea urchin. Sprinkle in diced spring onion and steam for a while. Serve.

◯◯ 零失敗技巧 ◯◯
Successful Cooking Skills

哪裏購買海膽？

一般大型超級市場或日式壽司店均有出售，價錢相宜。

Where to buy sea urchin?

It can be bought from large supermarkets or Japanese sushi stores with reasonable price.

如何蒸得嫩滑的蛋白？

火力勿太大，先用中火再調至慢火蒸熟；或蓋上耐熱保鮮紙，均可蒸成嫩滑的蛋白。

How to make smooth and soft steamed egg white?

The heat must not be high. Steam over medium heat first then turn to low heat and cook until done. Or cover the container with heat-resistant cling wrap.

海膽買回來後，需要特別處理嗎？

毋須特別處理，緊記放於雪櫃冷藏，取海膽時務必謹慎，因海膽容易斷爛。

Does the sea urchin require special handling after purchase?

No just remember to refrigerate it and take it carefully since it breaks easily.

陳皮豆豉醬蒸鮮鮑魚

Steamed Live Abalones in Dried Tangerine Peel and Black Bean Sauce

◎ 材料

活鮑魚仔 8 隻

陳皮豆豉醬半湯匙（做法請參考 P.36）

粟粉 1 茶匙

糖半茶匙

熟油 1 湯匙

◎ 做法

1. 陳皮豆豉醬用糖拌勻，備用。

2. 用刷去掉鮑魚表面的潺液，剪去鮑魚咀，洗淨，抹乾水分。

3. 用刀在鮑魚肉上劃十字花，抹上少許粟粉，排於蒸碟內，鋪上適量陳皮豆豉醬，隔水大火蒸 5 至 6 分鐘，澆上熟油，趁熱享用。

◎ Ingredients

8 live small abalones

1/2 tbsp dried tangerine peel and black bean sauce, refer to P. 36

1 tsp cornstarch

1/2 tsp sugar

1 tbsp cooked oil

◎ Method

1. Mix sugar with dried tangerine peel and black bean sauce. Stir well. Set aside.

2. Remove the slime on the feet of the abalones with a brush. Cut off the mouth pieces. Rinse well and wipe dry.

3. Make light crisscross incisions on the feet of the abalones. Lightly dust cornstarch on their feet. Arrange on a steaming plate. Spread some dried tangerine peel and black bean sauce on them. Steam over high heat for 5 to 6 minutes. Dribble hot cooked oil on them. Serve hot.

零失敗技巧
Successful Cooking Skills

為何陳皮豆豉醬與糖拌勻？
砂糖能中和陳皮的苦澀味及豆豉的鹹味，令醬汁甘香美味。

Why do you mix dried tangerine peel and black bean sauce with sugar?

The sugar helps neutralize the bitterness of the dried tangerine peel and the saltiness of the fermented black beans. The sauce tastes even better with sugar.

為何在鮑魚肉劃上花紋及抹上粟粉？
令鮑魚肉容易熟透，而且抹上粟粉能封鎖鮑魚的汁液，令肉質鮮嫩。

Why do you make crisscross incisions on the abalone and dust it with cornstarch?

The incisions speed up the steaming time. The cornstarch seals in the juice of the abalone, keeping its flesh moist and tender.

蒸 5 至 6 分鐘，熟透嗎？
絕對可以，太久令肉質過韌，浪費食材。

Can abalones be cooked properly in 5 to 6 minutes?

That's what it takes to cook them perfectly. Do not overcook them. Otherwise, the abalone will be tough.

香菇肉醬拌蒸茄子
Steamed Eggplant in Mushroom Meat Sauce

◯◯ 材料
茄子 12 兩
香菇肉醬半碗
葱絲適量
麻油 1 茶匙

◯◯ 做法
1. 茄子去蒂，開邊，隔水大火蒸 10 分鐘至全熟，取出茄子用叉子或撕成幼條。
2. 茄子幼條放於碟上，鋪上香菇肉醬，隔水大火蒸 5 分鐘，最後澆上麻油，放上葱絲，享用時拌勻。

◯◯ Ingredients
450 g eggplant
1/2 bowl mushroom meat sauce
finely chopped spring onion
1 tsp sesame oil

◯◯ Method
1. Cut off the stem of the eggplant. Cut into halves along the length. Steam over high heat for 10 minutes until done. Tear it into strips with a fork.
2. Arrange the strips of eggplant on a plate. Arrange the mushroom meat sauce on top. Steam over high heat for 5 minutes. Drizzle with sesame oil and put finely chopped spring onion on top. Stir well before serving.

香菇肉醬
Mushroom Meat Sauce

冬菇 1 兩
免治豬肉半斤
乾葱茸 4 湯匙
薑茸 1 湯匙
蒜茸 1 湯匙
米酒 1 湯匙
水半杯

⊙⊙ 調味料
老抽、麻油各 2 湯匙
鹽 2 茶匙
糖 1 茶匙
胡椒粉少許

⊙⊙ 芡汁（拌勻）
粟粉 2 茶匙
水 3 湯匙

⊙⊙ 做法
1. 冬菇去蒂、洗淨，用水浸軟，擠乾水分，切碎。
2. 燒熱鑊下油 3 湯匙，下乾葱茸、薑茸及蒜茸炒香，加入免治豬肉及冬菇粒炒香，潷酒，下調味料及水煮滾，轉小火煮 20 分鐘，下芡汁煮滾即成。

⊙⊙ **Ingredients**
38 g dried shiitake mushrooms
300 g ground pork
4 tbsp finely chopped shallot
1 tbsp grated ginger
1 tbsp grated garlic
1 tbsp rice wine
1/2 cup water

⊙⊙ **Seasoning**
2 tbsp dark soy sauce
2 tbsp sesame oil
2 tsp salt
1 tsp sugar
ground white pepper

⊙⊙ **Thickening glaze (mixed well)**
2 tsp cornstarch
3 tbsp water

⊙⊙ **Method**
1. Cut the stems off the mushrooms. Rinse well. Soak them in water until soft. Squeeze dry. Chop them finely.
2. Heat a wok and add 3 tbsp of oil. Stir fry shallot, ginger and garlic until fragrant. Put in the ground pork and shiitake mushrooms. Stir fry until fragrant. Sizzle with wine. Add seasoning and water. Bring to the boil. Turn to low heat and simmer for 20 minutes. Stir in thickening glaze. Bring to the boil again.

香菇肉醬拌蒸茄子

◎ 零失敗技巧 ◎
Successful Cooking Skills

茄子為何先蒸 10 分鐘？

蒸透後容易撕成幼條。

Why do you steam the eggplant for 10 minutes first?

It can be torn into strip a lot more easily after it has been steamed.

為何最後澆上麻油？

令整道餸更添香氣，而且油分足夠。

Why do you drizzle with sesame oil at last?

It adds an extra fragrance to the dish and also greases the eggplant nicely.

如何挑選做香菇肉醬的豬肉？

宜選 7 分瘦 3 分肥的豬踭肉製成免治豬肉，令醬料香濃、有嚼口。所有材料必須炒透，令油脂分泌均勻，香口惹味；香菇肉醬放入雪櫃可保鮮 3 天。

How do I choose the right cut of pork?

I prefer pork fore leg meat with about 30% fat. Ground it or finely chop it. It has the right mix of meat and grease to give the sauce a meaty flavour and a lovely chew. All ingredients should be fried till done so that the grease will be evenly distributed and the aroma is heightened. You can keep the mushroom ground pork sauce in the fridge for 3 days.

麵豉醬蒸小黃花
Steamed Small Yellow Croakers with Fermented Soybean Paste

◎ 材料

小黃花魚 12 兩
麵豉醬 1 湯匙
熟油 1 湯匙
薑絲 1 湯匙
葱絲適量

◎ 做法

1. 小黃花魚劏淨及處理妥當，洗淨，抹乾水分，排於蒸碟內，鋪上麵豉醬及薑絲。

2. 小黃花魚隔水用大火蒸 7 分鐘，隔去半份汁液，澆上熟油，用適量葱絲裝飾即可。

◎ Ingredients

450 g small yellow croakers
1 tbsp fermented soybean paste
1 tbsp hot oil
1 tbsp shredded ginger
shredded spring onions

◎ Method

1. Scale and gut the small yellow croakers. Wash thoroughly. Wipe them dry and arrange on a plate. Spread the fermented soybean paste and ginger on top.

2. Steam the fish over high heat for 7 minutes. Drain half of the steaming sauce from the plate. Drizzle hot oil on top. Garnish with the spring onions. Serve.

⓪ 零失敗技巧 ⓪
Successful Cooking Skills

小黃花魚經常有售嗎？
大型街市之鹹水魚檔經常有售。
Are small yellow croakers often available?
They are always available at sea-fish stalls in large markets.

小黃花與大黃花的肉質有何分別？
小黃花的肉質更嫩滑，適合日常蒸煮。
What is the difference between small and large yellow croakers in texture?
Small yellow croakers have smoother texture suitable for steaming.

麵豉醬的味道會太鹹嗎？
麵豉醬的用量不多，味道不會太鹹。若嫌偏鹹，可酌減麵豉醬份量。
Is fermented soybean paste a bit salty?
It is not salty as only a small amount is used. If you want it lighter, reduce the amount of it.

想吃得健康，可刪去最後澆熟油的步驟嗎？
澆上熟油後，魚肉更具光澤。若想走健康路線，建議選用橄欖油。
I want to eat healthy. Can we skip the last step of drizzling hot oil on top?
The fish are lustrous with hot oil on top. For healthy diet, you may use olive oil.

南瓜煮回鍋肉
Braised Pumpkin and Pork Belly

南瓜煮回鍋肉

⟨⟨ 材料
瘦腩肉半斤
南瓜 1 斤
蒜肉 4 粒（拍鬆）
豆豉 1 湯匙
薑 3 片

⟨⟨ 調味料
五香粉 1/4 茶匙
鹽 1 茶匙

⟨⟨ 做法

1. 腩肉洗淨，放入滾水內，加入薑片煮滾，轉中小火焓約 40 分鐘（至腩肉全熟），盛起，過冷河，切塊備用。

2. 南瓜去皮，去籽，洗淨，切厚塊。

3. 腩肉放入鑊內，白鑊用中小火煎至釋出少許油，下蒜肉及豆豉炒香，加入南瓜及水 1 1/2 杯煮滾，再煮約 15 分鐘，最後下調味料煮片刻至汁液濃稠即成。

⟨⟨ Ingredients
300 g lean pork belly
600 g pumpkin
4 cloves garlic (slightly crushed)
1 tbsp fermented black beans
3 slices ginger

⟨⟨ Seasoning
1/4 tsp five-spice powder
1 tsp salt

⟨⟨ Method

1. Rinse the pork belly. Put in boiling water. Add the ginger and bring to the boil. Turn to low-medium heat and blanch for about 40 minutes (until the pork belly is done). Rinse with cold water. Cut into pieces. Set aside.

2. Peel the pumpkin. Remove the seed and rinse. Cut into chunks.

3. Fry the pork belly without oil over low-medium heat until it releases a little oil. Add the garlic and fermented black beans. Stir-fry until aromatic. Put in the pumpkin and 1 1/2 cups of water. Bring to the boil. Cook for about 15 minutes. Add the seasoning and cook for a while until the sauce thickens. Serve.

◯◯ 零失敗技巧 ◯◯
Successful Cooking Skills

焓腩肉的水可倒掉嗎？
建議保留腩肉上湯，用以炒煮南瓜，令南瓜多一份肉香味！

Can we dump the stock from blanching the pork belly?

Reserve the stock to cook it with the pumpkin. It gives the pumpkin an additional aromatic flavor of meat!

麵豉柳梅煮白豆角
Pork Tenderloin and Long Beans in Soybean Paste

◎ 材料

豬柳梅 3 兩
白豆角 12 兩
麵豉醬 1 湯匙（剁碎豆粒）
蒜肉 4 粒（拍鬆）

◎ 醃料

生抽半湯匙
粟粉 1 茶匙
水 2 湯匙

◎ 做法

1. 柳梅洗淨，切片，加入醃料拌勻醃半小時。
2. 白豆角摘去頭尾兩端，切短度，洗淨備用。
3. 燒熱鑊下油 1 湯匙，下蒜肉爆香，加入肉片炒勻，下麵豉醬、水 3/4 杯及白豆角炒勻，煮約 8 分鐘至白豆角軟腍享用。

◎ Ingredients

113 g pork tenderloin
450 g light green long beans
1 tbsp soybean paste (beans chopped up)
4 cloves garlic (slightly crushed)

◎ Marinade

1/2 tbsp light soy sauce
1 tsp cornflour
2 tbsps water

◎ Method

1. Rinse and slice the tenderloin. Mix with the marinade and rest for 1/2 hour.
2. Remove both ends of the long beans. Cut into short sections. Rinse and set aside.
3. Heat up a wok. Add 1 tbsp of oil. Stir-fry the garlic until fragrant. Add the tenderloin and stir-fry. Put in the soybean paste, 3/4 cup of water and long beans. Stir-fry and mix well. Cook for about 8 minutes until the long beans turn soft. Serve.

◎ 零失敗技巧 ◎
Successful Cooking Skills

如何挑選白豆角？
挑選飽滿挺直、色澤嫩綠、新鮮無腐爛的白豆角。
How to select light green long bean?
Fresh long bean is plump, straight, and verdant. Do not choose the rotten one.

梅子汁白切肉
Poached Pork in Plum Sauce

材料
梅頭瘦肉 1 斤
小青瓜 2 個
薑 3 片

 梅子汁
梅子 2 粒（去核、剁碎）
黃砂糖 2 湯匙
蒜茸 1 茶匙
凍開水 4 湯匙
＊ 調勻

 做法
1. 梅頭瘦肉洗淨，放入加了薑片之滾水內煮滾，轉中小火焓約 40 分鐘，取出，用凍開水沖至涼，備用。
2. 小青瓜洗淨，切去頭尾兩端，切條備用。
3. 梅頭瘦肉切薄片，排於碟上，澆上梅子汁及伴青瓜食用。

 Ingredients
600 g pork butt
2 baby cucumbers
3 slices ginger

 Plum sauce
2 plums (cored and chopped up)
2 tbsp brown sugar
1 tsp finely chopped garlic
4 tbsp cold drinking water
* mixed well

 Method
1. Rinse the pork butt. Put in boiling water added with the ginger. Bring to the boil. Turn to low-medium heat and poach for about 40 minutes. Remove. Rinse with cold drinking water until it cools down. Set aside.
2. Rinse the cucumbers. Cut away both ends. Cut into strips. Set aside.
3. Thinly slice the pork butt. Arrange on a plate. Sprinkle with the plum sauce. Serve with the cucumbers.

零失敗技巧
Successful Cooking Skills

梅頭瘦肉可預早一天焓熟嗎？
當然可以！焓熟後冷藏凍吃，效果一流，更可省掉翌日的烹調工序。
Can we poach the pork butt 1 day in advance?
Yes, of course! Refrigerating the poached meat makes its taste fantastic when served cold. It also reduces the cooking steps next day.

香菇肉醬本菇煮鮮腐竹
Hon-shimeji Mushroom and Beancurd Sticks in Mushroom Meat Sauce

◎ 材料

鮮腐竹半斤
本菇 1 包
豆苗 4 兩
香菇肉醬 1/3 碗（做法請參考 P.46）
薑 4 片

◎ 調味料

蠔油 1 湯匙

◎ 芡汁（拌勻）

粟粉 1 茶匙
水 2 湯匙

◎ 做法

1. 本菇切去末端，洗淨，隔去水分。

2. 鮮腐竹切段，用水略沖，隔去水分。

3. 豆苗洗淨，放入滾水內灼熟，隔去水分，放於碟上備用。

4. 燒熱鑊下油 2 湯匙，下薑片及香菇肉醬爆香，加入鮮腐竹及本菇炒勻，下調味料及熱水半杯煮片刻，用芡汁煮滾埋芡，盛起澆於豆苗上。

◎ Ingredients

300 g fresh beancurd sticks
1 pack Hon-shimeji mushrooms
150 g pea sprouts
1/3 bowl mushroom meat sauce, refer to P.46
4 slices ginger

◎ Seasoning

1 tbsp oyster sauce

◎ Thickening glaze (mixed well)

1 tsp cornstarch
2 tbsp water

◎ Method

1. Cut off the roots of the Hon-shimeji mushrooms. Rinse and drain well.

2. Cut the beancurd sticks into short lengths. Rinse quickly in water. Drain.

3. Rinse the pea sprouts. Blanch them in boiling water briefly. Drain. Arrange on a serving plate.

4. Heat a wok and add 2 tbsp of oil. Stir fry ginger and the mushroom meat sauce. Add beancurd sticks and Hon-shimeji mushrooms. Stir well. Add seasoning and 1/2 cup of hot water. Cook briefly. Stir in the thickening glaze. Pour the resulting mixture over the bed of pea sprouts on the serving plate. Serve.

◯◯ 零失敗技巧 ◯◯
Successful Cooking Skills

下調味料及熱水後，再煮多久？

所有材料容易熟透，下調味料及熱水後再炒煮 3 分鐘，埋芡即成。

How long do you cook the mixture after adding seasoning and hot water?

All ingredients cook quickly. Just stir and cook them for 3 minutes after adding seasoning and hot water. Stir in thickening glaze and serve.

配鮮腐竹有何好處？

鮮腐竹能吸收醬汁的香味，食味更佳。

Why do you use fresh beancurd sticks in this recipe?

Fresh beancurd sticks pick up seasoning very well and they taste great.

芹蒜鹹菜煮門鱔骨

Conger-Pike Eel Bone with Chinese Celery, Baby Garlic and Pickled Mustard Green

◎ 材料

門鱔骨半斤
潮州鹹菜 4 兩
芹菜 1 棵
蒜仔 1 棵
薑 3 片

◎ 醃料

胡椒粉少許

◎ 調味料

魚露半湯匙
胡椒粉少許

◎ 做法

1. 門鱔魚骨洗淨，斬成小塊，下醃料拌勻。

2. 芹菜去鬚頭，摘去葉片，只要芹菜莖，洗淨，切段；蒜仔去鬚頭，洗淨，切段。

3. 潮洲鹹菜切絲，洗淨，擠乾水分。

4. 燒熱鑊下油 2 湯匙，下薑片爆香，放入門鱔骨略煎，加入鹹菜、滾水 1 1/2 杯煮 10 分鐘，下蒜仔、芹菜及調味料煮至滾即成。

◎ Ingredients

300 g conger-pike eel bone
150 g Chaochou pickled mustard green
1 stalk Chinese celery
1 stalk baby garlic
3 slices ginger

◎ Marinade

ground white pepper

◎ Seasoning

1/2 tbsp fish gravy
ground white pepper

◎ Method

1. Rinse the conger-pike eel bone. Chop into small pieces. Mix with the marinade.

2. Remove the root of the Chinese celery. Pick off the leaves. Only use the stems. Rinse the stems and cut into sections. Remove the root of the baby garlic. Rinse and cut into sections.

3. Shred the pickled mustard green. Rinse and squeeze water out.

4. Heat up a wok. Add 2 tbsp of oil. Stir-fry the ginger until aromatic. Roughly fry the conger-pike eel bone. Put in the pickled mustard green and 1 1/2 cups of boiling water. Cook for 10 minutes. Add the baby garlic, Chinese celery and seasoning. Bring to the boil to finish.

芹蒜鹹菜煮門鱔骨

◎ 零失敗技巧 ◎
Successful Cooking Skills

哪些是潮州鹹菜？

用潮州包心大芥菜醃製而成，少鹹少酸，質感爽脆，宜燜煮或熬湯。

Which is Chaochou style pickled vegetable?

It is made by pickling giant curled mustard. Mildly salty and sour, it is crunchy and suitable for stewed dishes or cooking in soup.

可以單售門鱔骨嗎？

通常連門鱔肉一併購買，一魚兩吃，亦可向相熟的魚檔查問，或有意外收穫。

Can we just buy the conger-pike eel bone?

It is often bought with the conger-pike eel meat for cooking in two ways. You can consult your regular fish stalls and may be amazed.

最後才放入蒜仔及芹菜煮滾，味道足以散發嗎？

蒜仔及芹菜味道香濃，略煮已足夠散發香氣。

Is the flavour enough by cooking baby garlic and Chinese celery in the final step?

Just a short cooking makes their strong aroma spread.

油 鹽 水 花 螺

Babylon Shells in Salted Water

◎ **材料**

鮮活花螺半斤
紅辣椒 2 隻
蒜肉 4 粒（拍鬆）
芫茜少許

◎ **調味料**

鹽 1 茶匙
油 1 湯匙

◎ **做法**

1. 花螺用清水洗淨外殼，備用。
2. 燒滾清水 1 1/2 杯，下蒜肉及調味料煮滾，放入花螺，見花螺的外奄離殼，即下紅辣椒略煮，上碟，以芫茜裝飾，趁熱享用。

◎ **Ingredients**

300 g live babylon shells
2 red chillies
4 cloves skinned garlic (crushed)
coriander

◎ **Seasoning**

1 tsp salt
1 tbsp oil

◎ **Method**

1. Rinse the babylon shells. Set aside.
2. Bring 1 1/2 cups of water to the boil. Add the garlic and seasoning. Bring to the boil. Put in the babylon shells. When the operculum separates from the shell, put in the red chillies and cook for a moment. Place on the plate. Decorate with coriander. Serve warm.

◯◯ 零失敗技巧 ◯◯
Successful Cooking Skills

如何挑選花螺？

中型的花螺鮮甜爽口；體型太大的肉質偏硬；太小的螺肉又不多。

How to select babylon shells?

Medium shells are sweet and crunchy; large shells are tough in the meat texture; and the small ones are not meaty.

買回來的花螺，需要吐沙嗎？

花螺不會藏有沙粒，毋須吐沙，只需洗淨外殼，保持衛生即可。

Is it necessary for the shells to spill sand grains?

Babylon shells contain no sand grains. Just rinse their shells to keep clean.

調味料份量足夠嗎？

絕對足夠！用油鹽水烹調，能吃到海鮮的原汁原味，毋須添加過多的調味料。

Is the seasoning enough?

Absolutely! We can taste the real flavour of seafood by cooking them in water with oil and salt. It is no need to add a lot of seasoning.

烹調螺肉有甚麼要注意之處？

花螺肉非常鮮味，但切勿久煮；否則花螺肉變硬，難以消化。

Is there anything that needs my attention when cooking Babylon shells?

Babylon shells are flavourful and delicious. But don't overcook them. Otherwise, the flesh will be rubbery and very difficult to chew.

油焗膏蟹
Fried Roe Crabs

◎ 材料
膏蟹 2 隻（約 1 斤 4 兩）
薑 8 片

◎ 蘸汁
薑絲 1 湯匙
大紅浙醋 1 小碟

◎ 做法
1. 膏蟹原隻洗淨外殼，放入冰格冷藏半小時。
2. 燒滾清水半鑊，膏蟹排於碟上，隔水大火蒸 12 分鐘，取出。
3. 燒熱鑊下油 3 湯匙，下薑片爆香，放入膏蟹（蟹蓋朝上）用小火煎焗約 5 分鐘，上碟，伴蘸汁食用。

油焗膏蟹

◎ Ingredients
2 female mud crabs (about 750 g)
8 slices ginger

◎ Dipping sauce
1 tbsp shredded ginger
1 small dish Chinese red vinegar

◎ Method
1. Rinse the whole female mud crabs. Keep in a freezer for 1/2 hour.
2. Bring 1/2 wok of water to the boil. Arrange the crabs on a plate and steam over high heat for 12 minutes. Remove.
3. Heat up the wok. Put in 3 tbsp of oil. Stir-fry the ginger until fragrant. Add the crabs with the inside of the shell upward. Fry over low heat for about 5 minutes. Serve with the dipping sauce.

◎ 零失敗技巧 ◎
Successful Cooking Skills

為何用膏蟹放入冰格冷藏？
將蟹膏冷藏至略硬，蒸蟹時蟹爪不會斷掉，保持蟹隻原整。

Why freeze the female mud crabs?
It is to make them firm so that their claws and legs are intact when steamed.

煎封膏蟹時，蟹蓋為何朝上放於鑊內？
以免蟹膏流瀉，緊記此技巧，別浪費蟹膏啊！

Why put the inside of the shell upward when frying?
This is to prevent the crab roe from outflowing. Bear this technique in mind and do not waste any crab roe!

膏蟹為何先蒸後煎？
煎封之目的是將薑油迫入膏蟹內，並非原隻煎熟，故必須先將膏蟹蒸熟。將薑油迫入鮮蟹，可保存膏蟹的鮮香，正是精華所在。

Why did you steam the crabs first before frying them?
The frying process is to infuse the crabs with ginger aroma in the oil. Frying itself isn't enough to cook the crabs through. That's why I steam them till done first and then fry them to flavour them with ginger-scented oil. That would preserve the umami of the crabs.

味噌煎龍脷柳
Fried Sole Fillet with Miso

材料
急凍龍脷柳 2 塊

醃料（調勻）
味噌 1 湯匙
米酒 2 湯匙

做法
1. 急凍龍脷柳解凍，洗淨，抹乾水分，切成大塊，用醃料塗勻醃 15 分鐘。
2. 抹去魚柳表面的醃料，放入熱油鑊內，用慢火煎至兩面金黃全熟，趁熱享用。

Ingredients
2 frozen sole fillets

Marinade (mixed well):
1 tbsp miso
2 tbsp rice wine

Method
1. Thaw the sole fillets. Rinse and wipe them dry. Cut into large pieces. Spread with the marinade and rest for 15 minutes.
2. Wipe the marinade away from the fillet surface. Fry in a wok of hot oil over low heat until both sides turn golden brown and are fully done. Serve hot.

◎ 零失敗技巧 ◎
Successful Cooking Skills

如何有效地解凍魚柳？

放於雪櫃的下層冷藏格自然解凍最恰當；若時間急趕，魚柳盛於保鮮袋後泡於水，可加速解凍。

How to thaw fillets effectively?

It is most appropriate to put the fillets into the lower compartment of a refrigerator to thaw naturally. If in a rush, put them in a Zip-lock bag and soak it in water to speed up the thawing process.

為何抹去魚柳的醃料？

以免用火慢煎魚柳時容易焦燶。

Why wipe the marinade of the fillet away?

It is to avoid the fillet scorching when frying over low heat.

除用龍脷柳外，還可用哪款魚柳？

銀鱈魚、白鱈魚、鯖魚也是不錯的選擇。

Can we use other fillets instead?

The good alternatives are black cod, seabass and mackerel.

葱段鹽煎魚
Fried Fish with Spring Onion and Salt

Ⓘ 材料
黃腳鱲 2 條（約 1 斤）
蔥 4 條
粗鹽 1 湯匙

Ⓘ 做法

1. 黃腳鱲處理妥當，洗淨，抹乾水分，在魚身兩面斜界三刀，抹上粗鹽，醃 2 小時。

2. 蔥摘取蔥白段，洗淨，瀝乾水分。

3. 用廚房紙抹乾魚身兩面，放入油鑊煎香一面，加入蔥白段，待鮮魚及蔥白煎至金黃香脆，上碟即成。

葱段鹽煎魚

Ⓘ Ingredients

2 yellowfin seabreams (about 600 g)
4 sprigs spring onion
1 tbsp coarse salt

Ⓘ Method

1. Scale and gill the yellowfin seabreams. Rinse and wipe them dry. Score the fish by making three slashes in both sides. Rub with the coarse salt. Marinate for 2 hours.

2. Pick the white part of the spring onion. Rinse and drain.

3. Wipe the fish dry with kitchen paper. Fry one side of the fish until fragrant. Add the white part of the spring onion. When the fish and spring onion turn golden and crispy, dish up to finish.

◎◎ 零失敗技巧 ◎◎
Successful Cooking Skills

怎樣可保持鮮味？

買回來的鮮魚用布吸乾血水，冷藏可保鮮味。

How do you keep the umami of the fish?

Make sure you wipe off the blood on the fish and keep it in the fridge.

為何在魚身兩面剠上幾刀？

令魚肉短時間煎透，以免時間太久破壞肉質，而且容易令鹽分滲入魚肉。

Why score both sides of the fish?

It is to make the fish fully cooked quickly. Long frying will damage the meat texture. It also allows the salt to infuse into the fish.

為何只摘取葱白部分？

近鬚頭的葱白部分，葱香味特濃，略炒更能散發香氣。

Why use just the white part of the spring onion?

The white part near the root is particularly fragrant. Slightly stir-frying it helps spread the aroma.

市場上哪款鮮魚適合使用？

牙帶、黑鱲、馬友、紅魚及黃花魚等，都是不錯之選擇。

What kind of fish in the market is suitable for this recipe?

Hairtail, black seabream, threadfin, red snapper, and yellow croaker are also good.

做法如此簡單？

沒錯！鮮活的海魚毋須用太多調味或醬料配搭，輕輕用鹽調味，吃出海魚之鮮味！

The cooking method is so simple?

Yes! Cooking fresh marine fish does not require a lot of seasoning or sauce. It tastes fresh with just a little salt!

豆渣蓮藕餅
Soybean Pulp and Lotus Root Patties

◎ 材料
豆渣 4 湯匙
蓮藕半斤
鯪魚膠 3 兩
蝦米 1 湯匙
雞蛋 1 個
乾葱 2 粒

◎ 調味料
胡椒粉少許
鹽半茶匙

◎ 做法
1. 蓮藕去皮，洗淨，刨幼絲。
2. 蝦米洗淨，切碎；乾葱去外衣，洗淨，切碎。
3. 所有材料及調味料盛於大碗內，拌勻，冷藏 1 小時。
4. 燒熱平底鑊，下油 2 湯匙，舀起一小團蓮藕漿放入鑊內，煎至一面呈金黃色，反轉再煎另一面，壓平成蓮藕餅，煎至兩面金黃色，隔油，上碟食用。

◎ Ingredients
4 tbsp soybean pulp
300 g lotus root
113 g dace paste
1 tbsp dried shrimps
1 egg
2 shallots

◎ Seasoning
ground white pepper
1/2 tsp salt

◎ Method
1. Skin the lotus root. Rinse and finely shred.
2. Rinse and finely chop the dried shrimps. Remove the skin of the shallots. Rinse and finely chop.
3. Put all the ingredients and the seasoning in a big bowl. Mix well. Refrigerate for 1 hour.
4. Heat a pan. Pour in 2 tbsp of oil. Spoon a small amount of the lotus root paste into the pan. Fry until a side turn golden brown. Flip it over to fry the other side. Flatten it into a patty. Fry until both sides turn golden brown. Drain. Serve hot.

豆渣

Soybean Pulp

外表： 淨白，呈碎粒狀；炒後呈金黃色。

製法： 擠壓豆漿後，餘下的豆渣放於雪櫃可冷藏 2 至 3 日；或用白鑊炒香後，易於儲存。

挑選： 自製或於日式超市專售豆腐的店子有袋裝出售。

豆渣蓮藕餅

Appearance:	Clean and white in the form of crushed grains; turning golden after stir-frying.
Method:	Obtained by pressing soy milk out. It can be preserved in a refrigerator for 2 to 3 days, or kept by stir-frying without oil.
Choice:	Homemade or bought in bags at the beancurd stalls in Japanese supermarkets.

◎ 零失敗技巧 ◎
Successful Cooking Skills

蓮藕絲等材料拌勻後，為何冷藏？
令所有材料調合得更黏實，以免散開。
Why refrigerate the mixture of shredded lotus root mixture?
To let all the ingredients stick firm.

甚麼是豆渣？
擠壓豆漿後，餘下的是豆渣。豆渣放在雪櫃內可保鮮 2 至 3 日，如用白鑊炒香，放入密封瓶內，約可保存一星期。
What exactly is soy dreg?
To make soymilk, soybeans are ground and pressed. After all soymilk is extracted, the solid residue is soy dreg. Soy dreg lasts in the fridge for 2 to 3 days. If you fry it in a dry wok first, and then keep it in an airtight container, it would last for a week.

加入豆渣，口感如何？
令煎出來的蓮藕餅更香脆。
How does it taste by adding soybean pulp?
The lotus root patty is additionally fragrant and crunchy.

這份量約可弄幾件？
約可製成 10 至 12 塊。
How many pieces can be made in this recipe?
It can make 10 to 12 pieces.

蓮藕用刀切成幼絲，可以嗎？
當然可以，但緊記切得幼細，口感細緻。
Can we shred the lotus root with a knife?
Yes, of course. But remember to finely shred to keep the silky texture.

西檸蜜糖煎鵪鶉
Fried Quails in Honey Lemon Sauce

⊕ 材料

冰鮮鵪鶉 3 隻
乾葱茸 2 茶匙
生粉 1 湯匙

⊕ 醃料

鹽及糖各半茶匙
生抽 2 茶匙
紹酒及生粉各 1 茶匙
胡椒粉少許

⊕ 調味料

檸檬汁 3 湯匙
蜜糖 1 1/2 湯匙
吉士粉 1 茶匙
水 1 湯匙
＊ 拌溶

⊕ 做法

1. 鵪鶉洗淨，去淨內臟，抹乾水分，
   開邊，用醃料醃 30 分鐘。
2. 鵪鶉輕輕地撲上生粉。
3. 燒熱適量油，放入鵪鶉半煎炸至金
   黃色，盛起。
4. 燒熱少許油，下乾葱茸爆香，放入
   調味料用慢火慢慢煮至濃稠，下鵪
   鶉兜炒至吸收汁料，上碟享用。

⊕ Ingredients

3 chilled quails
2 tsp finely chopped shallot
1 tbsp caltrop starch

⊕ Marinade

1/2 tsp salt
1/2 tsp sugar
2 tsp light soy sauce
1 tsp Shaoxing wine
1 tsp caltrop starch
ground white pepper

⊕ Seasoning

3 tbsps lemon juice
1 1/2 tbsps honey
1 tsp custard powder
1 tbsp water
* stir to solve

⊕ Method

1. Rinse and gut the quails. Wipe
   dry. Cut into halves. Mix with the
   marinade for 30 minutes.
2. Coat the quails thinly with the caltrop
   starch.
3. Heat up some oil. Fry and deep-fry
   the quails until golden. Set aside.
4. Heat up a little oil. Stir-fry the shallot
   until fragrant. Put in the seasoning
   and simmer until the sauce is thick.
   Add the quails and stir-fry until they
   absorb the sauce. Serve.

西檸蜜糖煎鵪鶉

◎ 零失敗技巧 ◎
Successful Cooking Skills

調味料加入了吉士粉，有何用途？

吉士粉具增香味、添色澤的效果，豐富檸檬蜜糖汁的香味，加添淡黃色澤，而且令調味汁更黏滑，賣相絕佳！

What is the use of adding custard powder in the seasoning?

Custard powder enhances flavour and colour. It makes the honey lemon sauce smelling rich with a yellowish tint, and carrying a smooth and sticky texture. The presentation is excellent!

家裏只有粟粉，鵪鶉可撲上粟粉代替生粉嗎？

鵪鶉撲上生粉油炸，令外皮更香脆，咬入口，無可抗拒！

There is only cornflour at home. Can it replace caltrop starch for coating the quail?

Caltrop starch will make the skin of the deep-fried quail crisper. You couldn't resist with just a bite!

鵪鶉很難購買嗎？

只有冰鮮或急凍鵪鶉出售，可預早幾天向商販訂購，或到大型凍肉店購買。

Is it hard to buy quails?

Only chilled or frozen quail is available. You can reserve from the vendors a few days earlier, or buy at large frozen food shops.

鼠尾草黑毛豬腩肉卷
Sage and Black Iberian Pork Belly Rolls

鼠尾草

鼠尾草黑毛豬腩肉卷

◎ 材料
黑毛豬腩肉薄片 10 塊
鼠尾草適量

◎ 蘸汁
紅椒碎半茶匙
蒜茸 1 茶匙
青檸汁 1 湯匙
魚露 1 湯匙
米醋 2 湯匙
＊ 調勻

◎ 做法
1. 鼠尾草摘取葉片，洗淨，抹乾水分，備用。
2. 黑毛豬腩片放入平底鑊內，煎至微黃全熟，上碟，用鼠尾草捲好黑毛豬腩片，伴上蘸汁食用。

◎ Ingredients
10 slices black Iberian pork belly
sage

◎ Dipping sauce
1/2 tsp chopped red chilli
1 tsp finely chopped garlic
1 tbsp lime juice
1 tbsp fish gravy
2 tbsp rice vinegar
* mixed well

◎ Method
1. Pick leaves from the sage. Rinse and wipe them dry. Set aside.
2. Fry the pork belly slices in a pan until light brown and fully cooked. Remove and put the pork belly slice on the sage and roll up. Serve with the dipping sauce.

1

2

3

◎◎ 零失敗技巧 ◎◎
Successful Cooking Skills

哪裏購買黑毛豬腩肉片？

黑毛豬腩肉片肉質細嫩，肉與脂肪的比例均勻，日式超市有售！

Where to buy sliced Black Iberian pork belly?

With a soft texture and an equal proportion of meat to fat, the pork belly can be bought at Japanese supermarkets.

即煎即吃嗎？

當然！熱辣香口，特別滋味！

Is it served right after fried?

Yes, of course! It is warm and spicy. How yummy!

有甚麼香草可代替鼠尾草？

鼠尾草帶淡淡的香草味，伴黑毛豬腩片最匹配。其他香草如羅勒、迷迭香等，香味較濃烈，不宜選用。

What herbs can be used instead of sage?

Sage has a light herbal aroma and goes well with the pork belly. Other herbs such as basil and rosemary have a heavy flavour which is not suitable for this recipe.

韓式辣醬拌牛肉
Beef in Korean Chili Sauce

材料

新鮮牛肉 4 兩（軟腍）
洋葱半個
黃芽白 6 兩
甘筍 2 兩
薑 4 片

調味料

韓式辣醬 2 茶匙

做法

1. 牛肉洗淨，切薄片，備用。

2. 黃芽白洗淨，切絲；洋葱去外衣，洗淨，切絲；甘筍去皮，洗淨，切絲。

3. 燒熱鑊下油 1 湯匙，下薑片及洋葱炒香，放入黃芽白、甘筍、水 3 湯匙及鹽半茶匙，炒至黃芽白軟身，加入牛肉及調味料炒片刻，至牛肉剛熟即可。

Ingredients

150 g fresh beef (soft)
1/2 onion
225 g Napa cabbage
75 g carrot
4 slices ginger

Seasoning

2 tsp Korean chili sauce

Method

1. Rinse and thinly slice the beef. Set aside.

2. Rinse and shred the cabbage. Peel, rinse and shred onion and carrot.

3. Heat up a wok. Add 1 tbsp of oil. Stir-fry the ginger and onion until aromatic. Put in the cabbage, carrot, 3 tbsp of water and 1/2 tsp of salt. Stir-fry the cabbage until soft. Add the beef and seasoning. Stir-fry for a while until the beef is just done.

零失敗技巧
Successful Cooking Skills

牛肉哪部位最軟腍？
牛肉最軟滑的部份是牛柳，肉質細緻，而且脂肪少。
Which part of beef is most tender?
It is the tenderloin. It has a delicate texture and less fat.

牛蒡甘筍炒牛肉絲
Stir-fried Beef with Burdock and Carrot

材料
鮮牛蒡 4 兩
甘筍 2 兩
冰鮮牛肉 4 兩（軟腍）
蒜肉 3 粒

白醋水
白醋 1 湯匙
水 4 杯
* 調勻

醃料
生抽半湯匙
胡椒粉少許
粟粉 1 茶匙

調味料
蠔油 1 湯匙
鹽 1/6 茶匙

做法
1. 牛肉洗淨，切絲，加入醃料拌勻醃半小時。
2. 鮮牛蒡用百潔布擦淨外皮，洗淨，切絲，放入白醋水內浸 20 分鐘，盛起，瀝乾水分（或浸至下鍋前，才瀝乾盛起）。（圖 1-2）
3. 甘筍洗淨，去皮、刨絲。
4. 燒熱鑊下油 1 湯匙，下蒜肉爆香，放入牛肉絲炒至轉色，盛起。（圖 3）
5. 原鑊加入水 3 湯匙，放入牛蒡及甘筍炒片刻，最後加入調味料及牛肉絲炒片刻即成。（圖 4-5）

Ingredients
150 g fresh burdock
75 g carrot
150 g chilled beef (soft)
3 cloves garlic

White vinegar solution
1 tbsp white vinegar
4 cups water
* mixed well

Marinade
1/2 tbsp light soy sauce
ground white pepper
1 tsp cornflour

Seasoning
1 tbsp oyster sauce
1/6 tsp salt

Method
1. Rinse and shred the beef. Mix with the marinade and rest for 1/2 hour.
2. Rub the skin of burdock with a scourer pad. Rinse and shred. Soak in the white vinegar solution for 20 minutes (or soak until it is ready for cooking). Remove and drain. (fig. 1-2)
3. Rinse, peel and shred the carrot.
4. Heat up a wok. Add 1 tbsp of oil. Stir-fry the garlic until fragrant. Put in the beef and stir-fry until it changes color. Dish up. (fig. 3)
5. Add 3 tbsp of water in the same wok. Put in the burdock and carrot. Stir-fry for a moment. Add the seasoning and beef. Stir-fry for a moment. Serve. (fig. 4-5)

◯◯ 零失敗技巧 ◯◯
Successful Cooking Skills

這道菜式如何少油烹調？

水炒的菜式是以水代油炒煮，減少吸油量。此餸最後用水炒煮牛蒡及甘筍，美味依然！

How to cook this dish with less oil?

Use water instead of oil to stir-fry to reduce the absorption of oil. It is still delicious by stir-frying burdock and carrot with water in the final step.

菠蘿雲耳炒柳梅

Stir-fried Tenderloin with Pineapple and Cloud Ear Fungus

◎ 材料
豬柳梅 6 兩
雲耳 1/3 兩
新鮮菠蘿 3 塊
小紅椒 2 隻
蒜肉 2 粒（略拍）

◎ 醃料
生抽半湯匙
胡椒粉少許
粟粉半茶匙

◎ 酸甜汁
白醋 2 湯匙
茄汁 3 湯匙
糖 2 湯匙
鹽半茶匙
水 3 湯匙
粟粉 1 茶匙
＊ 調勻

菠蘿雲耳炒柳梅

◎ 做法
1. 雲耳用水浸 1 小時，剪去硬蒂，洗淨，再放入滾水內浸 10 分鐘，瀝乾水分。
2. 柳梅洗淨，切片，下醃料拌勻醃半小時。
3. 菠蘿切小塊；紅椒去籽，洗淨，切絲。
4. 燒熱鑊下油 1 湯匙，下蒜肉爆香，放入柳梅及熱水 3 湯匙，炒至肉片熟透，下雲耳炒勻，傾入酸甜汁炒至滾，最後下菠蘿及紅椒絲拌炒，至汁液濃稠即成。

◎ Ingredients

225 g pork tenderloin
13 g dried cloud ear fungus
3 pieces fresh pineapple
2 small red chilies
2 cloves garlic (slightly crushed)

◎ Marinade

1/2 tbsp light soy sauce
ground white pepper
1/2 tsp cornflour

◎ Sweet and sour sauce

2 tbsp white vinegar
3 tbsp ketchup
2 tbsp sugar
1/2 tsp salt
3 tbsp water
1 tsp cornflour
* mixed well

◎ Method

1. Soak the dried cloud ear fungus in water for 1 hour. Remove the hard stalks. Rinse and soak in boiling water for 10 minutes. Drain well.

2. Rinse and slice the tenderloin. Mix with the marinade and rest for 1/2 hour.

3. Cut the pineapple into small pieces. Remove the seed of red chilies. Rinse and shred.

4. Heat up a wok. Add 1 tbsp of oil. Stir-fry the garlic until scented. Put in the tenderloin and 3 tbsp of water. Stir-fry until done. Add the dried cloud ear fungus and stir-fry. Pour in the sweet and sour sauce. Stir-fry and bring the sauce to the boil. Finally put in the pineapple and red chili. Stir-fry until the sauce thickens. Serve.

◎ 零失敗技巧 ◎
Successful Cooking Skills

為何選用柳梅？
奉行少油煮食的理念，建議選購偏瘦及軟腍的柳梅，蒸、炒皆宜。

Why use tenderloin?

This is to pursue healthy cooking by using less oil. Select tenderloin that is lean and soft for steaming or stir-frying.

蒜茸炒番薯葉
Stir-Fried Sweet Potato Leaves with Garlic

材料
嫩番薯葉 12 兩
蒜茸 1 湯匙

調味料
鹽 3/4 茶匙
糖半茶匙
水半杯

芡汁
番薯粉 / 半湯匙
水 4 湯匙
＊ 調勻、過濾

做法
1. 番薯葉洗淨，飛水，瀝乾水分。
2. 燒熱鑊下油 3 湯匙，下蒜茸炒香，加入番薯葉炒勻，下調味料煮片刻，最後埋薄芡即成。

Ingredients
450 g young sweet potato leaves
1 tbsp finely chopped garlic

Seasoning
3/4 tsp salt
1/2 tsp sugar
1/2 cup water

Thickening glaze
1/2 tbsp sweet potato starch
4 tbsp water
* mixed well and filtered

Method
1. Rinse the sweet potato leaves. Blanch and drain.
2. Heat up a wok. Put in 3 tbsps of oil. Stir-fry the garlic until fragrant. Add the sweet potato leaves and stir-fry evenly. Pour in the seasoning and cook for a while. Mix in the thickening glaze to serve.

◎ 零失敗技巧 ◎
Successful Cooking Skills

番薯葉容易購買嗎？

香薯葉是健康食材，粗纖維極多，有益腸道健康。很多街市均有番薯葉出售，宜選小葉片的番薯葉。

Is it easy to buy sweet potato leaves?

With abundant coarse fiber good for the intestine, sweet potato leaves are healthy food ingredients available in many markets. It is better to choose small-sized leaves.

家裏沒有番薯粉，怎辦？

以生粉或粟粉埋芡亦可。

There is no sweet potato starch at home. How to do?

Use caltrop starch or corn starch instead.

嫩番薯葉需要埋芡嗎？

由於番薯葉帶少許粗澀的口感，故建議用番薯粉埋芡，吃起來較順滑；但番薯苗則不用埋芡。

Do young sweet potato leaves need to be mixed with thickening glaze?

As they taste a bit tough and bitter, it is better to smooth them with sweet potato starch solution, but not for sweet potato shoots.

豬膶瘦肉浸辣椒葉
Chilli Leaves with Pork Liver and Pork in Soup

（◯）材料
辣椒葉半斤
豬膶 4 兩
瘦肉 4 兩
薑 4 片

（◯）調味料
鹽 1 茶匙

（◯）做法

1. 辣椒葉只摘取葉片，洗淨，瀝乾水分。

2. 豬膶及瘦肉分別洗淨，切片。

3. 煮滾清水 2 杯，放入薑片及瘦肉煮 5 分鐘，下豬膶煮滾，加入辣椒葉，轉慢火浸至豬膶熟透，最後下調味料拌勻即成。

豬膶瘦肉浸辣椒葉

（◯）Ingredients

300 g chilli leaves
150 g pork liver
150 g lean pork
4 slices ginger

（◯）Seasoning

1 tsp salt

（◯）Method

1. Pick the chilli leaves off the stems. Rinse the leaves and drain.

2. Rinse the pork liver and lean pork separately. Cut into slices.

3. Bring 2 cups of water to the boil. Put in the ginger and lean pork. Cook for 5 minutes. Add the pork liver. Bring to the boil. Put in the chilli leaves. Turn to low heat and cook until the pork liver is fully done. Season with the salt to finish.

◎◎ 零失敗技巧 ◎◎
Successful Cooking Skills

如何儲存辣椒葉？

辣椒葉吹乾，盛於戳上小孔的保鮮袋內，再用報紙包裹，放入膠袋內冷藏，可保存新鮮。

How to keep chilli leaves?

After air-drying the chilli leaves, put them in a pierced food storage bag which is then wrapped in newspaper, placed in a plastic bag, and refrigerated to keep fresh.

豬膶毋須調味醃製，會帶腥味嗎？

由於品嘗豬膶之鮮香味，故毋須醃味，以薑片同煮可去掉腥味；切薄片容易浸煮至熟透。

Does the pork liver smell fishy without marinade?

It is no need to marinate the pork liver in order to taste its original flavour. The fishy smell is removed by cooking with ginger. Finely sliced pork liver is easily cooked through.

辣椒葉的口感如何？

帶少許野菜粗韌的口感，纖維豐富。

What is the mouth-feel of chilli leaf?

Rich in fiber, chilli leaf tastes of wild vegetables — rough and chewy.

金蒜冬菜浸莧菜

Chinese Spinach with Preserved Tianjin Cabbage and Garlic in Soup

材料

莧菜 12 兩
蒜肉 8 粒
冬菜半湯匙
焓豬肉湯汁 2 杯

調味料

鹽半茶匙

做法

1. 莧菜摘去菜頭，洗淨，摘短度，瀝乾水分。
2. 冬菜用水略洗，洗掉砂粒，擠乾水分。
3. 燒熱鑊下油 2 湯匙，下蒜肉炒至金黃，傾入焓豬肉湯汁煮滾，放入莧菜煮滾，轉慢火浸煮 5 分鐘，下調味料及冬菜煮至微滾即成。

Ingredients

450 g Chinese spinach
8 cloves skinned garlic
1/2 tbsp preserved Tianjian cabbage
2 cups soup obtained from boiled pork

Seasoning

1/2 tsp salt

Method

1. Remove the head of the Chinese spinach. Rinse and pick into short sections. Drain.
2. Rinse the preserved Tianjian cabbage slightly. Rinse sand grains off. Squeeze water out.
3. Heat up a wok. Put in 2 tbsp of oil. Stir-fry the garlic until golden. Pour in the boiled pork soup. Bring to the boil. Put in the Chinese spinach. Bring to the boil. Turn to low heat and cook for 5 minutes. Add the seasoning and preserved Tianjian cabbage. Cook until the soup slightly boils to finish.

零失敗技巧 Successful Cooking Skills

沒有焓豬肉湯汁，怎辦？
改用盒裝上湯或清水代替，但香味降低不少。
There is no boiled pork soup. How to do?
Use packed stock or water instead, but the fragrance is greatly reduced.

容易洗掉冬菜的砂粒嗎？
砂粒容易藏於小葉片內，用水浸泡或放於水喉下沖洗，有效洗掉砂粒。
Is it easy to rinse sand grains off the Tianjian cabbage?
Soak it in water or rinse under running tap water. It can remove the sand grains inside the leaves effectively.

用慢火浸煮，莧菜會煮得太久嗎？
莧菜用慢火浸煮，可吸收肉汁精華，別擔心煮得過久。
The Chinese spinach is cooked over low heat. Will it be overcooked?
Don't worry! Cooking it gently over low heat helps take up the essence of meat sauce.

油鹽水浸油鱲仔
Pony Fishes in Salted Water

材料

油鱲仔半斤
芫茜 4 棵
葱 2 條
薑 4 片

調味料

海鹽 3/4 茶匙

做法

1. 油鱲仔劏好，洗淨，瀝乾水分。
2. 芫茜及葱去掉鬚頭，洗淨。
3. 燒熱鑊下油 1 湯匙，下薑片爆香，傾入清水 1 1/2 杯煮滾，放入油鱲仔煮至微滾浸片刻，最後下調味料、芫茜及葱煮滾即成。

Ingredients

300 g pony fishes
4 stalks coriander
2 sprigs spring onion
4 slices ginger

Seasoning

3/4 tsp sea salt

Method

1. Scale and gill the pony fishes. Rinse and drain.
2. Remove the root of the coriander and spring onion. Rinse.
3. Heat up a wok. Put in 1 tbsp of oil. Stir-fry the ginger until scented. Pour in 1 1/2 cups of water. Bring to the boil. Add the pony fishes. Cook until the water is slightly boiled. Soak for a while. Put in the seasoning, coriander and spring onion. Bring to the boil to finish.

零失敗技巧
Successful Cooking Skills

用油鹽水浸煮油鑶仔,是最佳吃法嗎?

油鑶仔是海魚的上品,魚味鮮美、肉質嫩滑,用油鹽水烹調可保持其原汁原味。

Is it the best way to taste pony fishes by cooking in salted water with oil?

Pony fishes are superb marine fish. Their original flavour and smooth texture are retained through cooking in this way.

需要大量芫茜、薑及葱浸煮嗎?

天然的薑葱香味,令海魚的味道得到全面提升。

Need to use a generous amount of coriander, ginger and spring onion?

The natural aromatic flavour of ginger and spring onion brings the fish to a higher level of taste.

如何令湯底更鮮甜味香?

若喜歡的話,可加入大頭菜或冬菜等一併浸煮,令湯底更香甜,多一層味道體驗。

How to make the soup base sweeter and more fragrant?

If you like preserved turnip or preserved cabbage, cook with it to add the depth of flavour to the soup, making it more delicious.

十穀米釀鮮魷筒
Baked Stuffed Squid with Ten-grain Rice

ⓘ 材料

鮮魷 1 隻（12 兩，約 25 厘米長）
十穀米 3 兩
蒜茸 1 茶匙
清水 225 毫升

ⓘ 醃汁

鰻魚汁 2 湯匙
黑胡椒碎 1 茶匙
蒜茸半湯匙
麻油少許

ⓘ 調味料

鹽 1/3 茶匙

ⓘ 蘸汁

水 2 湯匙
生粉半茶匙
餘下的醃汁

ⓘ 做法

1. 鮮魷撕去外衣，取出鮮魷鬚及軟骨，洗淨，抹乾魷魚筒內外，保持圓筒狀。

2. 醃汁拌勻，均勻地塗抹在鮮魷兩面，醃 1 小時。

3. 十穀米洗淨，用清水浸 2 小時，加入調味料拌勻煮熟，待涼。

4. 預熱焗爐 200℃。

5. 十穀米飯釀入鮮魷筒內至 7 至 8成滿，用牙籤封口。

6. 焗盤鋪上錫紙，放上鮮魷筒焗 8分鐘，取出，塗上餘下之醃汁，再焗 8 分鐘，切件，上碟。

7. 下油爆香蒜茸，加入蘸汁材料煮滾，伴鮮魷筒享用。

十穀米釀鮮魷筒

ⓘ Ingredients

1 fresh squid (450 g and about 25 cm long)
113 g ten-grain rice
1 tsp minced garlic
225 ml water

ⓘ Marinade

2 tbsp eel sauce
1 tsp chopped black pepper
1/2 tbsp minced garlic
sesame oil

ⓘ Seasoning

1/3 tsp salt

ⓘ Dipping Sauce

2 tbsp water
1/2 tsp caltrop starch
remaining marinade sauce

◎ Method

1. Tease outer skin from fresh squid and remove tentacles and soft bones. Rinse and wipe out its inside and outside. Keep its cylindrical shape.

2. Mix the marinade and rub over the two sides of fresh squid evenly. Set aside for 1 hour.

3. Rinse ten-grain rice and soak in water for 2 hours. Mix in seasoning and cook until done. Set aside to let cool.

4. Pre-heat an oven to 200°C.

5. Stuff ten-grain rice into fresh squid until medium-well full and fix the end with toothpick.

6. Lay aluminum foil over a baking tray. Put in the squid and bake for 8 minutes. Rub over the remaining marinade and bake for 8 minutes more. Cut into pieces and put into a plate.

7. Heat oil in a wok and stir-fry minced garlic until fragrant. Add the ingredients of the dipping sauce and bring to the boil. Serve with fresh squid at the side.

◎ 零失敗技巧 ◎
Successful Cooking Skills

為何十穀米加鹽煮熟？
令十穀米帶淡淡味道，不致寡淡無味。
Why cooking ten-grain rice with salt?
This makes the ten-grain rice has a light flavor.

釀鮮魷筒有何秘訣？
將十穀米釀入鮮魷筒時，米飯不可鬆散，否則切件時，飯粒容易散開。
What's the tip of stuffing fillings into fresh squid?
The ten-grain rice to be stuffed into fresh squid cannot be scattered or loosed or else the rice scatters when the squid is cut into pieces.

麻醬凍豆腐
Cold Beancurd in Sesame Sauce

材料
盒裝嫩豆腐 1 盒
乾海藻 1 湯匙

調味料（拌勻）
麻醬 2 湯匙
生抽半湯匙
蘋果醋 2 茶匙
麻油 2 茶匙
糖 1 茶匙

做法
1. 乾海藻用清水浸泡約 5 分鐘，用凍開水略沖，瀝乾水分。
2. 盒裝嫩豆腐盛起，隔去水分，鋪上海藻，澆上調味料即可享用。

Ingredients
1 pack silken beancurd
1 tbsp dried seaweed

Seasoning (mixed well):
2 tbsp sesame sauce
1/2 tbsp light soy sauce
2 tsp apple cider vinegar
2 tsp sesame oil
1 tsp sugar

Method
1. Soak the dried seaweed in water for about 5 minutes. Slightly rinse under cold drinking water. Drain.
2. Dish the beancurd up. Drain. Place the seaweed on top. Sprinkle with the seasoning and serve.

零失敗技巧
Successful Cooking Skills

應挑選哪款盒裝豆腐？
建議選絹豆腐，質感滑嫩，冷吃最佳。
What kind of packed beancurd should we choose?
The best choice is silken beancurd which has a smooth and silky texture. It is a perfect cold starter.

調味料為何加入蘋果醋？
令醬汁帶微酸甜，口感豐富。
Why season with apple cider vinegar?
The vinegar gives the source slightly sweet and sour tastes, enriching the flavour of the dish.

哪裏購買乾海藻？
於大型日式超市購買，包裝出售，價錢略貴。
Where can we buy dried seaweed?
A bit costly, it can be available at large Japanese supermarkets in packets.

紅燜木耳豬腱肉
Braised Pork Shoulder with Wood Ear Fungus

材料

豬腱肉 1 斤（連筋）
木耳 1 兩
薑 8 片
紹酒 1 湯匙

 調味料

老抽 1 1/2 湯匙
冰糖 1 湯匙
鹽半茶匙

 做法

1. 木耳用水浸透，剪去硬蒂，洗淨，撕成小塊，飛水，過冷河，備用。

2. 豬腱肉洗淨，切大塊，飛水，過冷河備用。

3. 燒熱鑊下油 1 湯匙，下薑片爆香，放入豬腱，灒酒炒勻，傾入水 3 杯煮滾，轉小火燜 45 分鐘，加入木耳及調味料煮滾，再轉小火燜半小時，至豬腱軟腍及汁液濃稠即可。

 Ingredients

600 g pork shoulder (with tendons)
38 g dried wood ear fungus
8 slices ginger
1 tbsp Shaoxing wine

 Seasoning

1 1/2 tbsp dark soy sauce
1 tbsp rock sugar
1/2 tsp salt

 Method

1. Soak the dried wood ear fungus in water. Remove the hard stalks. Rinse and tear into small pieces. Blanch and rinse with cold water. Set aside.

2. Rinse the pork shoulder. Cut into chunks. Blanc and rinse with cold water. Set aside.

3. Heat up a wok. Add 1 tbsp of oil. Stir-fry the ginger until fragrant. Put in the pork shoulder. Sprinkle with the Shaoxing wine. Stir-fry and mix well. Pour in 3 cups of water and bring to the boil. Turn to low heat and cook for 45 minutes. Add the dried wood ear fungus and seasoning. Bring to the boil. Turn to low heat and cook for 1/2 hour until the pork shoulder turns tender and the sauce thickens. Serve.

零失敗技巧
Successful Cooking Skills

為何選用豬腱肉？
豬腱肉肥膏不多，肉中帶筋特別好吃，飛水後可去掉過多的油分。

Why choose pork shoulder?

It contains less fat and the tendons attached to the meat make it chewy and delicious. Blanch the meat to remove excessive oil.

零失敗
秘方系列

啖啖鮮香
住家餸
The best of homestyle cooking

編者
Forms Kitchen編輯委員會

Editor
Editorial Committee, Forms Kitchen

美術設計
馮景蕊

Design
Carol Fung

排版
劉葉青

Typography
Rosemary Liu

出版者

Publisher
Forms Kitchen

香港鰂魚涌英皇道1065號
東達中心1305室
電話
傳真
電郵
網址

Room 1305, Eastern Centre, 1065 King's Road,
Quarry Bay, Hong Kong.
Tel: 2564 7511
Fax: 2565 5539
Email: info@wanlibk.com
Web Site: http://www.wanlibk.com
            http://www.facebook.com/wanlibk

發行者

Distributor

香港聯合書刊物流有限公司
香港新界大埔汀麗路36號
中華商務印刷大廈3字樓
電話
傳真
電郵

SUP Publishing Logistics (HK) Ltd.
3/F., C&C Building, 36 Ting Lai Road,
Tai Po, N.T., Hong Kong
Tel: 2150 2100
Fax: 2407 3062
Email: info@suplogistics.com.hk

承印者

Printer

中華商務彩色印刷有限公司

C & C Offset Printing Co., Ltd.

出版日期
二零一九年五月第一次印刷

Publishing Date
First print in May 2019

鳴謝以下作者提供食譜（排名不分先後）：
黃美鳳、Feliz Chan、Winnie姐